YOUR KNOWLEDGE HAS VALUE

- We will publish your bachelor's and master's thesis, essays and papers

- Your own eBook and book - sold worldwide in all relevant shops

- Earn money with each sale

Upload your text at www.GRIN.com and publish for free

Axel Stelter

Hanseatic Architecture

GRIN Verlag

Bibliografische Information der Deutschen Nationalbibliothek:

Die Deutsche Bibliothek verzeichnet diese Publikation in der Deutschen National-
bibliografie; detaillierte bibliografische Daten sind im Internet über http://dnb.d-
nb.de/ abrufbar.

Imprint:

Copyright © 2008 GRIN Verlag GmbH
Druck und Bindung: Books on Demand GmbH, Norderstedt Germany
ISBN: 978-3-640-24497-3

Axel Stelter

Architecture 170B

GSI: Ocean Howell

November 2008, UCB

The Hanseatic League's Influence on the Architecture of the Baltic Area

In 1894 the British architect John Tavenor wrote an article about the remains of medieval architecture in the Baltic area. He concluded that the style in this area has been carried throughout the Middle Ages and further stated that the style is "quite dissimilar to those of the rest of the continent"[i], meaning the Gothic style, that started to spread over Europe in the 11th and 12th centuries. He calls this architecture the "Baltic style" and considers it to be a sub-style of the Gothic style.

The Holy Roman Empire, which contributed to the spread of the Gothic style, only reached as far as the Elbe River in Germany. The Baltic area, as the heart of Northern Europe, was fractured into many kingdoms, principalities and lordships in the 1st millennium B.C. So how was it possible that cultural and economic goods could spread in these disadvantageous circumstances, at a time when "commerce by sea was little more than outrageous piracy and commerce by land was obliged to follow one or two beaten tracks across Europe in order to escape merciless exactions of the robber barons"[ii]?

One answer could be the Hanseatic League, a protected network created by merchants, in order to protect their trade. This alliance allowed trading guilds to manifest a trade monopoly within the entire Baltic area. Since the League was not tied to any sphere of control but the merchants themselves, trades could be made easily within Northern Europe.

Consequently the simultaneous appearance of the League and the Baltic style suggest that there is a possible correlation between the architecture in the Baltic area and its spread along the Hanseatic League's trading routes. During this essay I am going to support this assumption by finding exemplifying similarities among buildings in the Hanse towns along the main trading routes.

In order to show a possible correlation between the style of architecture and the trading routes of the League it is important to compare towns and their architecture at each end of that route because they might offer similarities. Many historians see the town of Lübeck in Northern Germany as the mother or "queen of the Hanse", because the year of its foundation is also considered the beginning of the Hanseatic League era. Tavenor goes as far as to say that "the history of Lübeck becomes the history of the Hansa".[iii] That means that most of the main trading routes have their origin here, which suggests using Lübeck as the starting point in tracing the Baltic style. The Hanse town Lübeck was founded in the year 1143 and soon after, sources prove the appearance of a German merchant organization, a forerunner of the Hanseatic League.[iv] The town is located at the southern end of the Baltic Sea and has direct connections to the European mainland; hence its central location is ideal for sea and land trade.

Lübeck and Rostock's Old Town Halls (see "Example 1" in appendix)

The first example that implies a correlation between the style of architecture and the trading routes of the Hanseatic Leagues are façade and gables of the so-called *kontors*, the branch offices. Although one expects them to be more or less alike in the different areas of the League, because they were functional trading buildings, the *kontors* offer a clear impression of how the League spread the same style over the sea.

A *kontor* in the context of the Hanseatic League is a either a foreign trading post, which usually included a warehouse and further facilities for the merchants to operate their businesses abroad, or simply the house of a merchant. Most of time the merchants incorporated their entire life in the *kontor*. The warehouse, office (*kontor* is also an old expression for the word "office"[v]) and optionally a shop were united under one roof with the merchant's family.

The most distinct part of these contour buildings in the hanseatic era carried the typical gable style, which generally characterized the term of the "Baltic style". A gable is the upper enclosing wall surface of a building in the area roof area. It is also considered to be short for the term gable wall, which defines the entire outer wall from the gable down to the ground. These gables can provide information about of where the Hanseatic League exported its style along its trading routes in the Baltic.

Because Lübeck is the mother of all Hanse towns it makes sense to find the first example for a contour and its gable here. There are many different beautiful and unique gables in the old town center but one stands out the most. The town hall building was originally built in 1150 and served as a building for the merchants of the League. Since then it has exemplified a mélange of the architectural styles affecting Lübeck in the last centuries.[vi]

The larger gable, added in the 13th century, consists of two main walls is enclosed by three octagonal towers with copper flags. Each wall has three arched blind windows, each of which contains two decorative windows and one circle. Two round and decorative holes in the wall reveal that there is no actual building mass behind this gable. The whole gable appears like a defensive wall of a castle, but its height also emphasizes the importance and influence of the building beneath it. Monochrome brick served as a material for the gable itself. Copper was used as roofing for the walls and the towers and made the gable shine and express power.

The Hanse town of Rostock joined the Hanseatic League in 1283.[vii] It is located about 150 kilometers along the coast east of Lübeck. Because Rostock could be reached from Lübeck within less than just a day by ship, it was a frequently used destination. Just like Lübeck, Rostock built a town hall along the market in the 13th century, which also became the center of communication between the hanseatic merchants.[viii] It served as a court, storage and shopping facility. Today the medieval aspects of the brick Gothic are mostly covered by other styles. But as in Lübeck, the gable still stands out and allows room to find similarities between the two original buildings at the time of the Hanseatic League.

The original building from 1218 is a part of two middleclass houses, which are connected by a gabled roof. Nowadays the front façade facing the market expresses a mixture of Baroque, and Renaissance additions, but the gothic gable on top still reaches out into the sky. The gable wall has seven towers reaching up, where the middle tower is the highest. Each tower has a small copper flag on top of it and is shaped octagonal. Every part of the gable is roofed with copper and a long cornus reaches horizontally to from one side to the other between the outside towers. The gable wall features a total of twelve decorated arches, which are symmetrically distributed. The extended vertical axes of two adjacent towers enclose two arches. Each arch consists of two symmetric blind windows and an ornamental circle on top. Two different colors of brick additionally decorate the overall appearance of the gable wall through usage of alternating rows of red and black bricks.

Conclusion

The functions of the town halls in Rostock and Lübeck in the 12th and 13th century and especially their appearances are very similar. Of course, in this case the two town halls were important functional buildings for the Hanseatic League and it seems obvious that they should look similar. But although there was no higher power mandating a certain style or building program other than the town itself, the two town halls' appearance share a lot of similarities, where the most striking similarities are the symmetry of the gable wall with towers, the material and the incorporation of decorative arches with blind windows. Conclusively the trading activities of the Hanseatic League from Lübeck to Rostock strongly influenced the decisions made about the town hall in Rostock approximately 50 years later.

St.Mary's in Lübeck and St.Nikolai in Stralsund (see "Example 2" in appendix)

Besides spreading similarities in trading buildings which provide economically effective functions, the Hanseatic League also spread a style in sacral buildings. Finding similarities in religious buildings proved a deeper cultural exchange along the trading routes beyond the reason of commerce.

Again, it is of advantage to start the investigation in Lübeck. Because of its optimal location the town received the seat of the bishopric[ix] which led to the building of churches and cathedrals. St. Mary's was a milestone regarding its size and structure. The church "became the prototype for many main churches on both sides of the Baltic as far as Riga and Reval".[x] It influenced other cities of the Hanseatic League to emulate it; hence relevant building information had to be transferred via trading routes.

A closer investigation of St. Mary's in the town of the "Queen of the Hanseatic League" reveals that the church was first built in with brick in Romanesque style in 1156. With Lübeck's growing importance as a trading center, the size was not sufficient anymore and the leaders of the town wanted to express the economic and political power of Lübeck in the new design. Further, it was intended to build a structure which symbolized the desire of freedom for the merchants. In 1250 the new construction started. Influenced by Gothic cathedrals in France, the Lübeckians adapted their structure of the hall and used a system of buttress to deduct the weight from the high towers. The use of brick instead of natural stone was caused by the geological composition of the area: there simply was no rock in the ground. St. Mary's was the first church of its kind in the Northern European Baltic area because of its height achieved only by brick construction. The inner brick vault still counts as the highest in world.[xi]

The original re-design of this church was built as a three aisled basilica[xii] with Lübeck brick instead of the initially planned ashlars stone. All the roofing for the towers (twin towers substituted the old single tower) and nave was copper and the side aisles were roofed with brick tiles. Externally the church did not have "any of the decoration which ashlars-Gothic had developed"[xiii], but typical brick façade decorations like blind windows and arch shaped ornaments. Further the front façade is strongly symmetric and the sites of each of the hexagonal shaped towers look exactly the same. The nave is significantly higher than the side aisles and its buttress system reaches over the roof structure of the side aisle (called "flying buttresses"). The church also includes two separately roofed apsidal choirs, which stand out further and higher than the side aisle in order to create the cruciform plan. The apse consists of three semi-circular shapes to form one circular vaulted extension.

Stralsund at the Southern end of the Baltic Sea is only about a one day trip by ship apart from the Hanse town Lübeck, which allowed a frequent exchange of goods. Although the amount of trading was significantly less compared to other big Hanse towns in the 13th century, Stralsund was strongly influenced by the Hanseatic League.

As late as 1294 the town of Stralsund officially joined the Hanseatic League; the Church of St. Nikolai in Stralsund was built in 1276. Originally it was a hall church but in 1314 the city decided to change it into a basilica with two towers. The original roofs of the two towers were originally quadrangular and not very steep. The façade of the towers looked exactly the same on all four sides, of which each a three blind windows (some windows were real) on each level. Materials for the building were mainly brick. Copper was used for each roof. A small tower on top of the nave reaches out above the center of the cruciform shape of the church. The nave's flying buttress system was extended over the significantly lower aisles. The apse is shaped semi-circularly.

Conclusion

In conclusion, St. Mary's and St.Nicolai are very much alike in all parameters, as size, layout, material and façade. Already the most apparent parameters like the flying buttress system, the twin towers and the similarities of naves and aisles indicate their common origin. The fact that the building in Stralsund was only built a few years later after St. Mary's and was also re-designed leads one to believe that the trading of the Hanseatic League had a major influence in their resemblance. Since Lübeck and Stralsund were two different city governments, untied to no shared power base, a connection by a political force can be excluded. A transmission of the style via land trade is also very implausible because several principalities, like Rostock, Wismar and Schwerin[xiv] along the 250 kilometer way would have had to be crossed, each of which could add customs duties to travelers and goods. Hence the sea trade, under the name of the Hanseatic League, must have had an impact on the transfer of the style of the architecture.

St. Maria in the non-hanseatic town Güstrow (see "Example 3" in appendix)

The previous examples have shown how the styles of sacral buildings and league-related commercial buildings have impacted Hanse towns along the trading routes. By the end of the 13th century there were about 50 cities along the Baltic Sea that were members of the League However, towns that did not have the privilege of being a part of the Hanseatic League were not as influenced by the spread of the Baltic Style.[xv] These mostly were towns with no accessible waterways and could not use the advantage of the trade by ship and hence had to stay excluded. This exclusion emphasizes the League's impact on spreading architecture: non-League settlements did not exhibit as many architectural elements of the Baltic Style. This last example of a sacral building during the high time of the Hanseatic League will show how towns, which were not involved in the league's trading activities, have a dissimilar architecture.

The dome in Güstrow is considered as a building that expresses the transition from the Roman style to the Gothic style.[xvi] This already shows the general belatedness of the architecture in this relatively isolated town, since Lübeck's St. Mary's was built at the same time and counts as Gothic example. It was built in the first half of the 13th century and the main material is brick. Brick tiles were used for all roofing, except for the small copper-covered tower in the center of the cruciform shape. Some structural pillars are made of granite. The west end of the church is a single tower with a regular pitched roof. Each of the horizontal sections has a different amount of blind windows and the sections are separated by a cornus. The dome's cruciform shaped is achieved by a transept which is located in the center of the whole structure. Only one aisle is located on the east-south end, but the west end has two symmetric aisles. All aisles are structurally disconnected from the nave and are divided into separate small sections of which each has its own pitched roof.

Conclusion

Comparing this Roman-Gothic dome to St. Mary's or St. Nikolai offers many obvious differences. Lübeck and Stralsund exhibit a strong symmetric building with twin towers and slim, high-reaching peeks, whereas Güstrow's dome has a more massive and heavy appearance. Also the difference in the building technology is noticeable: Güstrow does not yet use the flying buttress system, nor does the dome connect the roof structure of the aisles to the nave. Similarly as in Lübeck and Stralsund, the dome in Güstrow uses the typical blind windows and brick art work to beautify the façade. Conclusively, St. Maria incorporates elements that are very similar to sacral buildings in that part of Europe during the 13th century but the overall building does not express any attempt of incorporating elements of the League's sacral building architecture. The lack of influence, caused by the town's isolation from the Hanseatic League, must be a reason for the dissimilarity of Güstrow's dome.

Conclusion

These carefully chosen examples have shown that the Hanseatic League facilitated the transmission of the Baltic Style from one Hanse town to another. Both sacral buildings and League-related buildings exhibit certain characteristics which are closely related. However, I believe that towns excluded from the League were not able to participate as strongly in the trade and hence their architecture was not as strongly influenced by the League. Of course in the grand scheme, there were other factors contributing to the spread of the Baltic Style, which were not mentioned here such as religious Christian movements towards the East, non-League related trade and tribal wars. During the Middle Age in Northern Europe they also influenced the spread of culture and hence specific architecture. In any case, the Hanseatic League was the only uniting factor in the Baltic area during the 12th and 13th centuries because Europe was divided into many pieces. Consequently the birth and growth of the Hanseatic League facilitated the exchange of architectural and cultural elements between towns and countries.

Appendix

Example 1

Townhall Lübeck 1960's[xvii]

Main Gable, Townhall Lübeck[xviii]

13th Century Townhall Rostock[xix]

Townhall Rostock[xx]

Townhall Rostock, 2002[xxi]

Example 2

St. Mary's and Town Hall in Lübeck, 14th century[xxii]

St. Mary's Lübeck, 1960's[xxiii]

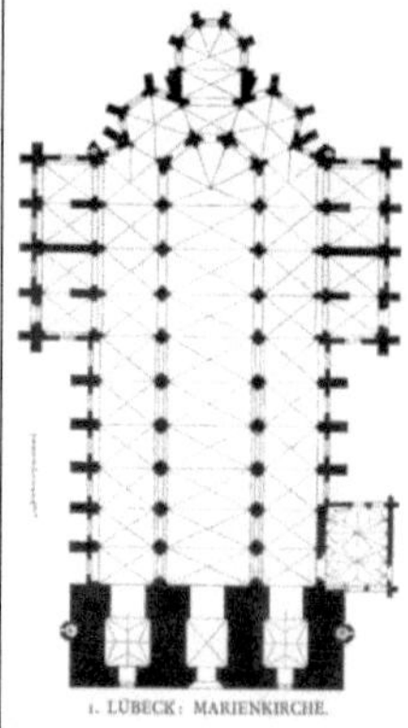

Floorplan St. Mary's[xxiv]

St. Mary's 1890's[xxv]

St. Nikolai, Stralsund

Flying Butresses, St Nikolai[xxvi]

St. Nikolai from the East[xxvii]

Example 3

Dome in Güstrow, approx. 2000[xxviii]

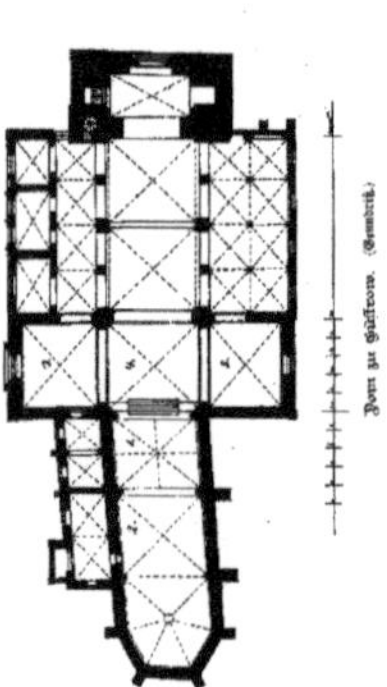

Floorplan, Güstrow's Dome[xxix]

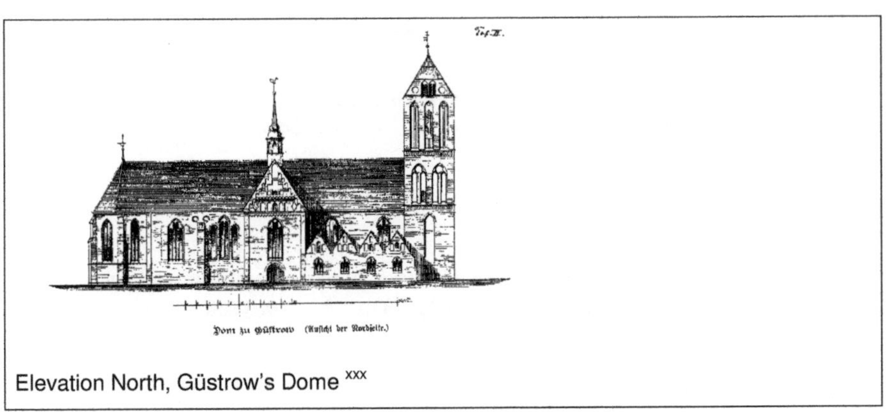

Elevation North, Güstrow's Dome [xxx]

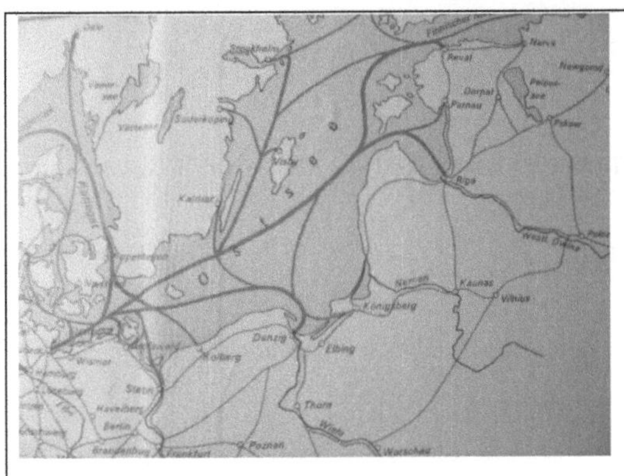

Map of Trading Routes from 1250-1450[xxxi]

References

[i]Perry, John Tavenor, *Influence of the Hanseatic League on the architecture of Northern Europe*, Royal Institute of British Architects. Journal, 1893-94, 3rd ser., v. 1, p. 473-493, 1893,
Page 475

[ii]Blackall, Clarence Howard, *Hanseatic architecture*, American architect and building news, 1886 Mar. 6, v. 19, p. 112-116 ; 1886 Apr. 3, p. 159-160, 1886,
Page 159

[iii] Perry, John Tavenor, *Influence of the Hanseatic League on the architecture of Northern Europe*, Royal Institute of British Architects. Journal, 1893-94, 3rd ser., v. 1, p. 473-493, 1893,
Page 478

[iv]Enns, A. B., Lübeck; *A guide to the architecture and art treasures of the Hanseatic town*, Lübeck 1965,
Page 5

[v]Konrad Fritze, Johannes Schildhauer, Walter Stark, *Die Geschichte der Hanse*, GDR 1974,
Page 145

[vi]Enns, A. B., Lübeck; *A guide to the architecture and art treasures of the Hanseatic town*, Lübeck 1965,
Page 15

[vii]Perry, John Tavenor, *Influence of the Hanseatic League on the architecture of Northern Europe*, Royal Institute of British Architects. Journal, 1893-94, 3rd ser., v. 1, p. 473-493, 1893,
Page 492

[viii]Schröder, Jan, *Der mittelalterliche Ursprungsbau*, in: *Das Rostocker Rathaus*, Rostock, 2002 ,
Page 4-8

[ix]Enns, A. B., Lübeck; *A guide to the architecture and art treasures of the Hanseatic town*, Lübeck 1965,
Page 16

[x]:Enns, A. B., Lübeck; *A guide to the architecture and art treasures of the Hanseatic town*, Lübeck 1965,
Page 22

[xi]: Visited 11/21/2008
http://www.luebeck.de/tourismus/sightseeing/sehenswuerdigkeiten/objekte/marien.html

[xii] Enns, A. B., Lübeck; *A guide to the architecture and art treasures of the Hanseatic town*, Lübeck 1965,
Page 29

[xiii] Enns, A. B., Lübeck; *A guide to the architecture and art treasures of the Hanseatic town*, Lübeck 1965,
Page 30

[xv] Konrad Fritze, Johannes Schildhauer, Walter Stark, *Die Geschichte der Hanse*, GDR 1974,
Page 1 (map)

[xvi] Visited 11/21/2008
http://dlib.uni-rostock.de/servlets/YearbookInquiry?docid=3271

<u>Bibliography</u>

Zur hanseatischen Kunst des Mittelalters , Hartlaub, Gustav Friedrich,

Influence of the Hanseatic League on the architecture of Northern Europe, Perry, John Tavenor

Hanseatic architecture, Blackall, Clarence Howard.

Die Hanse : Aufstieg, Blutezeit und Niedergang der ersten europaischen Wirtschaftsgemeinschaft : eine Kulturgeschichte von Handel und Wandel zwischen 13. und 17. Jahrhundert, By Uwe Ziegler

Die Hanse, By Klaus Friedland

England and the German Hanse, 1157-1611 : a study of their trade and commercial diplomacy, By T.H. Lloyd

Die Geschichte der Hanse, By Konrad Fritze, Johannes Schildhauer, Walter Stark

Lübeck; a guide to the architecture and art treasures of the Hanseatic town, By A. B. Enns

German Hansa, by N. Dollinger

http://dlib.uni-rostock.de/servlets/YearbookInquiry?docid=3271

http://www.luebeck.de/tourismus/sightseeing/sehenswuerdigkeiten/objekte/marie n.html

http://www.stralsund.de

http://www.rostock.de

http://www.barlachstadtguestrow.de

http://www.nikolai-stralsund.de/frame.html

References

[xvii] Enns, A. B., *Lübeck; A guide to the architecture and art treasures of the Hanseatic town*, Lübeck 1965,
Page 34

[xviii] Visited 23/11/08:
http://static.panoramio.com/photos/original/7221024.jpg

[xix] Visited 20/11/2008:
http://upload.wikimedia.org/wikipedia/commons/9/99/VSR-Rathaus.jpg

[xxi]Visited 20/11/2008:
www.rostock.de/rathaus/pics

[xxii] Konrad Fritze, Johannes Schildhauer, Walter Stark, *Die Geschichte der Hanse*, GDR 1974,
Page 47,48, 50

[xxiii] Enns, A. B., *Lübeck; A guide to the architecture and art treasures of the Hanseatic town*, Lübeck 1965,
Page 5

[xxiv] *Kirchliche Baukunst des Abendlandes,* Georg Dehio,Gustav von Bezold
Buchhandlung 1887-1901, Plate No. 448.

[xxv] Perry, John Tavenor, *Influence of the Hanseatic League on the architecture of Northern Europe*, Royal Institute of British Architects. Journal, 1893-94, 3rd ser., v. 1, p. 473-493, 1893,
Page 481

[xxvi] Visited 23/11/08:
http://content.answers.com/main/content/wp/en-commons/thumb/b/b0/180px-Stralsund,_st_Nikolai_(2007-01-24)_d.JPG

[xxvii] Visited 23/11/08:
http://www.hanse.org/files/seiteninhalt/hansestaedte/stralsund/

[xxviii] Visited 23/11/08:
http://farm4.static.flickr.com/3263/2555777374_28886e978e.jpg?v=0

[xxix] Visited 23/11/08:
http://dlib.uni-rostock.de/servlets/YearbookInquiry?docid=3271

[xxx] Visited 23/11/08:
http://dlib.uni-rostock.de/servlets/YearbookInquiry?docid=3271

[xxxi] Konrad Fritze, Johannes Schildhauer, Walter Stark, *Die Geschichte der Hanse*, GDR 1974,
Page 1 (map)